AF594782

River Monsters

Golden Dorado

MARYSA STORM

Bolt is published by Black Rabbit Books
P.O. Box 227, Mankato, Minnesota, 56002
www.blackrabbitbooks.com

Alissa Thielges, editor; Rhea Magaro, designer and photo researcher

Library of Congress Cataloging-in-Publication Data
Names: Storm, Marysa, author.
Title: Golden dorado / by Marysa Storm.
Description: Mankato, Minnesota: Black Rabbit Books, [2024] | Series: River monsters | Includes bibliographical references and index. | Audience: Ages 8-12 | Audience: Grades 4-6 | Summary: "With jaws as strong as bolt cutters, golden dorado are powerful predatory fish. Reel in hi-lo readers with leveled text and infographics about their features, South American habitat, and hunting strategies"—Provided by publisher.
Identifiers: LCCN 2023035684 (print) | LCCN 2023035685 (ebook) | ISBN 9781623109127 (library binding) | ISBN 9781623109288 (ebook)
Subjects: LCSH: Salminus brasiliensis—Juvenile literature.
Classification: LCC QL638.B88 S76 2024 (print) | LCC QL638.B88 (ebook) | DDC 597/.48—dc23/eng/20231117
LC record available at https://lccn.loc.gov/2023035684
LC ebook record available at https://lccn.loc.gov/2023035685

Printed in China.

Image Credits

Alamy: Franco Banfí cover, Jose Luis Suerte 13 (t), 25, 32, Larry Larsen 9, Luke Massey 18, puzzle M 6–7, WaterFrame_fba 4–5, 20–21; MISSING: 8–9, 19 (bl); Shutterstock: AlexLMX 12, Aneta Jungerova 13 (bl), Couperfield 13 (br), Curioso. Photography 20 (river), Iurii Vlasenko 29 (br), Ivo Antonie de Rooij 20 (falls), Jarun Ontakrai 26 (inset), Jose Luis Stephens 26, McCarraher's Photo 0p 21 (trees), Olga Popova 22, OutdoorWorks 23, Patife 10–11, 31, Renato Zaar 3, Rudmer Zwerver 19 (bm), TR STOK 19 (br), Triple R Publishing House 1, 14–15, 16–17 (bkgd), 19 (br), 28–29

Contents

CHAPTER 1

Mysterious Beasts

Beneath the surface of the Paraná River, a little sabalo swims. A storm brews in the sky above. Strong winds whip the water. The little fish is distracted by the change in weather. It doesn't notice the golden dorado **lurking** behind a log.

Attack!

The golden dorado has been waiting for the perfect moment to feed. And now it's arrived. In a flash, the **predator** lurches forward. Its mighty jaws snap around the smaller fish. Crunch! It's time to eat.

The word *dorado* means "golden" in Spanish.

CHAPTER 2

Size and Strength

Golden dorados are powerful predators. These river fish are often about 3 feet (0.9 meters) long but can get much bigger. They grow up to 66 pounds (30 kilograms). The females are often larger than males. One of the largest golden dorados ever caught weighed the same as a Labrador retriever.

Average Size

0
5
10
15
20
25
30
35
40
45
50
pounds
pounds
WEIGHT
25
POUNDS
(11.3 kg)

PARTS OF GOLDEN DORADO

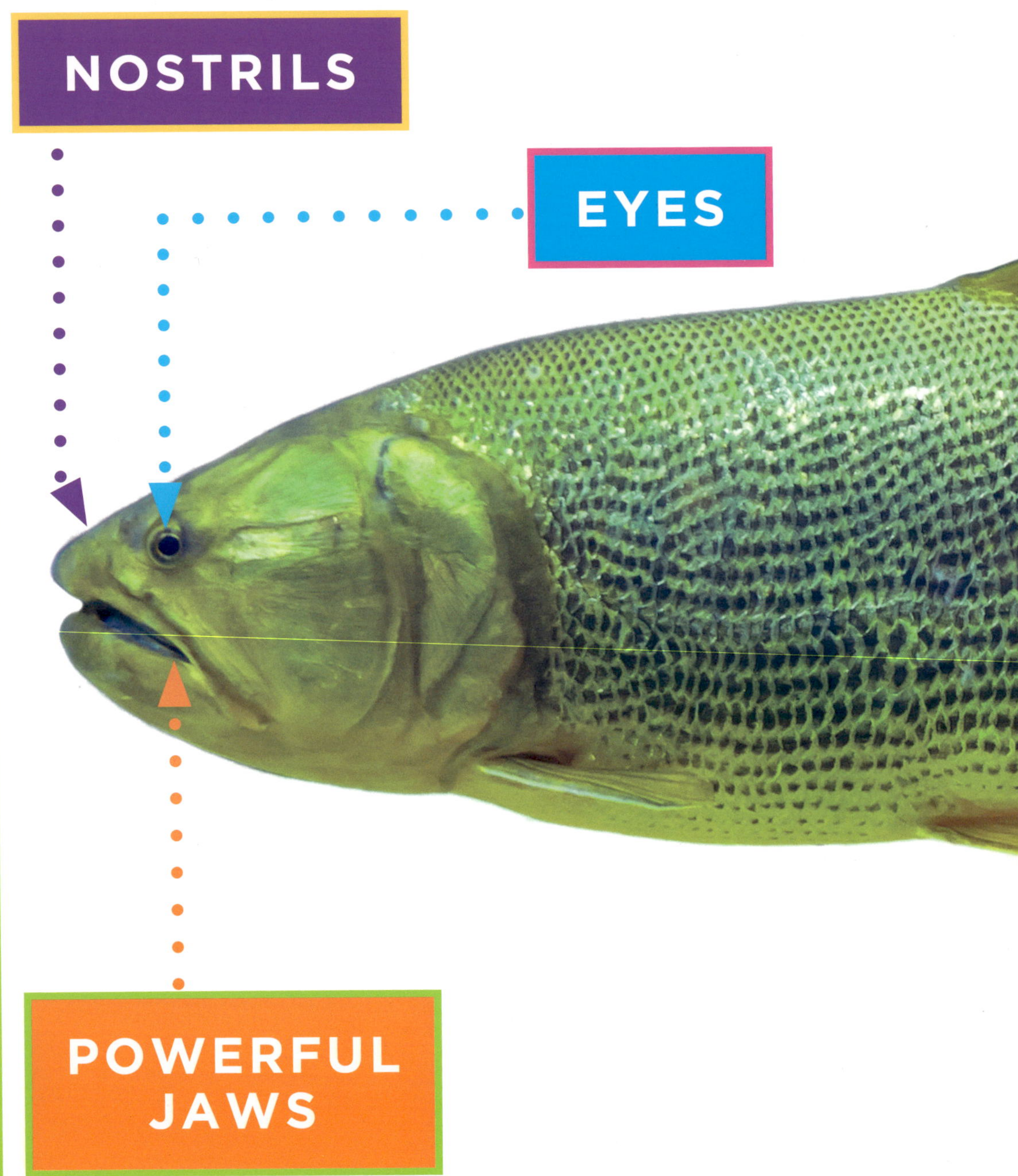

FINS
GOLD-COLORED SCALES
TAIL

Black and Gold

This fish gets its name from its gold scales. Its sides are often flecked with black. It has an orange tail. Some tails have a black stripe down the center. These fish have large mouths, strong jaws, and tiny, sharp teeth.

How STRONG is a GOLDEN DORADO'S Jaw?

pit bull jaws

bolt cutters

CHAPTER 3

Hunting Habits

Their life span is roughly 15 years.

Golden dorados are freshwater fish. They swim, feed, and **spawn** in South America. People find them in Brazil, Paraguay, Uruguay, Bolivia, and Argentina. They **migrate** between late October and January to breed.

WHERE GOLDEN DORADOS LIVE

SOUTH AMERICA

BRAZIL
BOLIVIA
PARAGUAY
URAGUAY
ARGENTINA

Top Predators

Adult golden dorados are apex predators. Other animals may eat young or injured ones. But a big, healthy dorado is hard to catch. Golden dorados mostly eat other fish. They also catch lizards, mice, and small birds.

giant otter

Golden Dorado Food Chain

This food chain shows what golden dorados eat.

GOLDEN DORADO

WHERE THEY LIKE TO HUNT

Beneath fallen
trees

On the Hunt

These fish sometimes hunt and travel in groups. When they find fish to eat, it can be a **feeding frenzy**. Golden dorados are patient hunters. They usually wait until their **prey** is distracted to strike.

Female golden dorados lay many eggs. Few live to become adults.

CHAPTER 4

Dangers and Threats

The golden dorado is a challenging fish to catch. When hooked, it puts up a big fight. It leaps into the air and jerks at the line. These traits make it a prized catch. Many people travel to South America to fish for it.

Most people catch and release dorados. They don't keep the fish once it's caught.

EL
KARAÍ

These fish can have trouble with parasites.

Learning More

Golden dorados aren't in danger of dying out. But **developments** like dams are problems. These structures may make it harder for them to migrate.

Thankfully, people try to keep their habitats safe. There are groups studying this fish too. There's still a lot to learn about this incredible creature.

By the Numbers

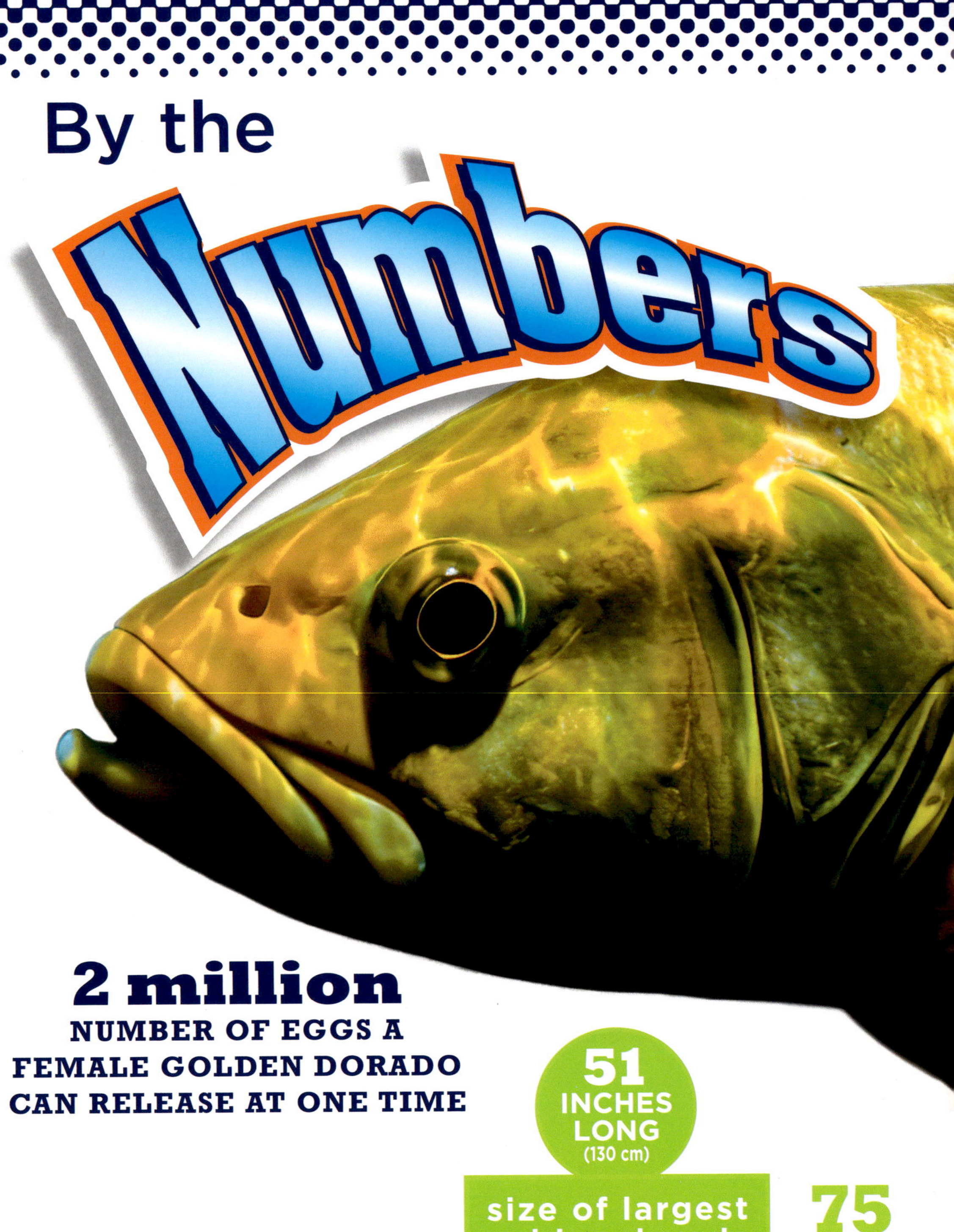

2 million
NUMBER OF EGGS A FEMALE GOLDEN DORADO CAN RELEASE AT ONE TIME

51
INCHES LONG
(130 cm)

size of largest golden dorado ever caught

75
POUNDS
(34 kg)

400 TO 700

miles (644 to 1,127 kilometers)
HOW FAR SOME GOLDEN DORADO SWIM TO SPAWN

3 FEET (0.9 m)

how high they can leap in the air

68 TO 82 DEGREES FAHRENHEIT (20 to 28 degrees Celsius)

PREFERRED TEMPERATURE OF WATER TO SWIM IN

GLOSSARY

development (dih-VEL-uhp-muhnt)—an area of land with buildings that were all built around the same time

feeding frenzy (FEE-ding FREN-zee)—a state of wild activity in which the animals in a group are all trying to eat something

lurk (LURK)—to secretly wait in a place to do something harmful

migrate (MY-grayt)—to move from one place to another

parasite (PAR-uh-syt)—a plant or animal that lives in or on another plant or animal and causes harm

predator (PRED-uh-tuhr)—an animal that eats other animals

prey (PRAY)—an animal hunted or killed for food

spawn (SPAWN)—to produce or deposit a large number of eggs

BOOKS

Doyle, Abby Badach. *Freshwater Fishing.* New York: Gareth Stevens, 2023.

Mazzarella, Kerri. *Fly Fishing.* New York: Crabtree Publishing, 2023.

Reeves, Diane Lindsey. *Freshwater Fishing.* Minneapolis: Lerner Publishing, 2024.

WEBSITES

Dorado
www.pantanalescapes.com/wildlife/fish/dorado.html

Fly Fishing Facts for Kids
kids.kiddle.co/Fly_fishing

Meet the 'Lightning Fast, Super Lacerater,' Golden Dorado
www.youtube.com/watch?v=xyTBhQ4LwEE

INDEX